Tryptophan

Powerful Serotonin Booster

Barbara Wexler, MPH

WOODLAND PUBLISHING

For permissions, ordering information, or bulk quantity discounts, contact:
Woodland Publishing, 448 East 800 North, Orem, Utah 84097

Visit our Web site: www.woodlandpublishing.com
Toll-free number: (800) 777-BOOK (2665)

The information in this book is for educational purposes only and is not recommended as a means of diagnosing or treating an illness. All matters concerning physical and mental health should be supervised by a health practitioner knowledgeable in treating that particular condition. Neither the publisher nor the author directly or indirectly dispenses medical advice, nor do they prescribe any remedies or assume any responsibility for those who choose to treat themselves.

Cataloging-in-Publication data is available from the Library of Congress.

ISBN 978-1-58054-121-3

Printed in the United States of America

Contents

Psst . . . Got L-Tryptophan?

Chances are you've heard of L-tryptophan, the biologically active form of the amino acid tryptophan. Unfortunately, the first time many people learned about tryptophan was when it made headlines in 1989. We'll get to that twisted tale in a moment, but first, let's find out why tryptophan is "essential" and what it does in the body.

Tryptophan is found in many animal and plant proteins and is considered essential because it is not produced by the human body; we have to get tryptophan, along with eight other essential amino acids, from outside sources—from our diet or from a nutritional supplement.

Tryptophan is a precursor, or building block, that the body uses to make 5-hydroxytryptophan (5-HTP), an aromatic amino acid that is a precursor of serotonin, an important neurotransmitter. Neurotransmitters are naturally occurring chemicals in the brain that transmit messages from one nerve cell to another. Serotonin is involved in thought, emotions and behavior, and in the production of melatonin, a neurohormone associated with regulating sleep-wake cycles.

The Tale of the Sleepy Turkey Eaters

But you're probably most familiar with the urban legends and lore surrounding tryptophan. For example, it is generally believed that the drowsiness we experience after eating our traditional Thanksgiving feast is caused by the tryptophan found in turkey. Like most urban legends, this one begins with a kernel of truth—tryptophan does produce a mild sedative effect because it is readily converted into serotonin in the body, and turkey does contain tryptophan.

But here's where we have to counter this intuitively appealing notion with fact, because the truth is that eating turkey alone—even on an empty stomach, which is optimal for obtaining the effects of tryptophan—is unlikely to produce any measurable effect, let alone the post-holiday-dinner urge to nap. That's because the quantities of tryptophan in turkey, even in unusually large portions, are far too low to exert such an effect. So there's probably not much chance of falling into a tryptophan-induced stupor after eating turkey. You're better off blaming chronic sleep deprivation, low blood sugar or the tedium of an especially slow day for your mid-afternoon slump, because the turkey you just ate is not the culprit.

So what causes that lazy, lethargic feeling and urge to nod off right after Thanksgiving dinner? It is most likely the quantity of food ingested, and the immediate demands of digesting all of those carbohydrates, fats and protein. Not only is the body calling upon all of its resources for the purposes of digestion, there is also an increased demand for insulin, the hormone that controls blood sugar levels and enables the cells to use glucose for energy.

As you might imagine, the insulin demands of a meal so rich in carbohydrates are considerable. Whoever dreamed up a dinner menu so heavily skewed toward carbohydrates—cornbread, cranberries, mashed potatoes, candied sweet potatoes and bread stuffing, all before rich, sugary desserts—must have anticipated it would require all of our energy to digest it!

Of course, there are other factors that may contribute to the post-meal slump. Alcohol consumption may cause drowsiness, especially if you don't drink regularly. In the case of the cook, the exhaustion from preparing the feast may suddenly hit at the end of the meal. And Thanksgiving dinner guests may use fatigue as an excuse to avoid helping with the dishes!

Is Warm Milk a Magic Sleep Potion?

Did your mother serve you a cup of warm milk as you got ready for bed? Has warm milk endured as part of the traditional bedtime recipe for sound sleep because it contains tryptophan and melatonin?

Maybe, but many scientists and nutritionists believe that the small amounts of these substances in a glass of milk are not likely to shorten the time it takes to fall asleep or to improve the quality and duration of sleep. So why does warm milk seem to help soothe some people, especially children, and promote sleep? Most likely, the effect is as psychological as it is physical.

The chemical that may explain some of warm milk's sleep-inducing effects is probably tryptophan (roughly 112 milligrams per 8-ounce glass), which, along with the calcium and carbohydrates in milk, increases serotonin and melatonin production.

The milk's physical properties—such as its warmth—might also help too, since warmer temperatures tend to lull people to sleep. For many people, warm milk may prompt comforting memories of being soothed to sleep as an infant. Babies often fall asleep while nursing or drinking formula from a bottle, so it makes sense that as adults we might retain the association of milk and sleep—not cold milk from the refrigerator, but milk warmed to body temperature. People who respond favorably to warm milk may be making the association that warm milk will enable them to sleep like a baby.

The Role of Tryptophan in Human Health

Tryptophan is an organic compound, one of 20 amino acids commonly found in animal sources of protein. It is formed from proteins during digestion by the action of proteolytic enzymes, substances produced by the pancreas that break ingested proteins down into peptides or amino acids.

Tryptophan is necessary for the normal growth and development of infants and for nitrogen balance in adults. Nitrogen balance refers to the relationship between nitrogen taken into the body, usually as food, and nitrogen excreted from the body. A negative nitrogen balance means that some parts of the body, such as muscles, are breaking down faster than they are replaced.

Supplemental tryptophan is available as L-tryptophan, the biologically active form of the amino acid. It is also important to understand that for humans, only L-tryptophan is effective; the D-isomer, or acetyl-D form, of tryptophan is not useful.

L-tryptophan has been used to treat sleep problems, such as insomnia, and to improve the quality and duration of sleep. It also has been used to aid relaxation and to relieve symptoms of anxiety, depression and menstrual problems.

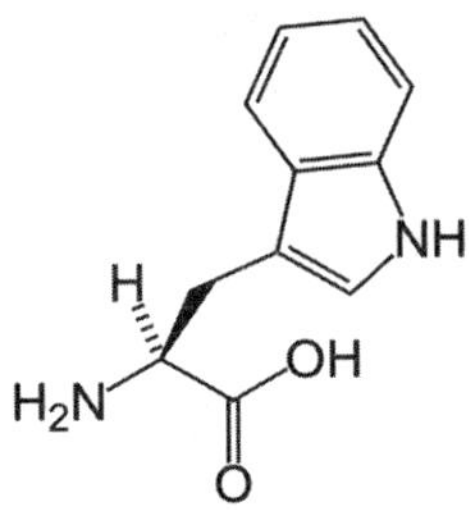

L-Tryptophan

L-tryptophan also appears helpful for relieving migraine headaches and may assist with weight management, smoking cessation and attention deficit disorder.

In addition, tryptophan is necessary for the production of niacin. Also known as vitamin B3, niacin supports healthy skin, nerves and digestion.

Tryptophan Deficiency

Since tryptophan is necessary for the production of niacin, tryptophan deficiency can increase the risk of niacin deficiency. A serious tryptophan deficiency may lead to pellagra. Symptoms of pellagra—known as the disease of the Ds—include diarrhea, dermatitis (a scaly rash), dementia (loss of intellectual and cognitive ability), depression and death. The origin of the name "pellagra" is Italian—*pelle* means "skin" and *agra* means "rough"—and describes one of the principal symptoms of the disease.

Tryptophan deficiency is most likely to occur in individuals with sharply decreased protein consumption and may lead to depression, insomnia, schizophrenia, suicidal thoughts and even carbohydrate cravings (important for dieters). In concert with magnesium deficiency, tryptophan deficiency may be a contributing factor for coronary artery spasms.

Today, pellagra occurs only rarely in populations that enjoy balanced diets, but it was once a major threat to public health in the United States. In 1927, after the Mississippi River flooded 16 million acres in seven states, there was a spike in the already high incidence of pellagra along the Mississippi River.

Dr. Joseph Goldberger, a physician working in the U.S. Public Health Service (at that time, the Hygienic Laboratory, which was the forerunner of the National Institutes of Health), alienated many members of the medical establishment when his research revealed that pellagra was not caused by an infectious agent, as was commonly believed at the time, but by a nutritional deficiency. Goldberger had the Red Cross add brew-

Dr. Joseph Goldberger

er's yeast (which provides quality protein, thiamin, riboflavin, iron and phosphorus) to food rations. In a few weeks, people suffering from pellagra recovered, and scores of new cases were prevented.

Goldberger's work also shed light on the relationship between poverty and pellagra. In addition to implicating the flood conditions in the outbreak, Goldberger observed that the agricultural system did not support healthy diets for sharecroppers, tenant farmers and mill workers. His work conclusively demonstrated that disadvantaged populations were most likely to have deficient diets and therefore were at high risk for pellagra.

Tryptophan Requirements

As mentioned earlier, tryptophan is an essential amino acid that we must acquire from our diets or from nutritional supplements. According to *The Merck Manual for Healthcare Professionals,* the weight-adjusted requirement for amino acids correlates with growth rate, which decreases from infancy until adulthood. For tryptophan, the requirement is 19 milligrams per kilogram of body weight (mg/kg) for infants 4–6 months old, 4 mg/kg for children 10–12 years old, and 3 mg/kg for adults. The World Health Organization recommends 3.5 to 4.5 mg/kg.

For those of you who have resisted learning the metric system, here are a couple of examples with kilograms converted into pounds—a small child of about 20 pounds (about 9 kilograms) needs about 112.5 milligrams of tryptophan per day, while a 155-pound (70-kilogram) adult needs about 245 milligrams.

Therapeutic dosages of L-tryptophan are generally higher and do vary; however, one 500 mg capsule of L-tryptophan for every 50 pounds of body weight is typically taken one hour before bedtime. To treat depression, adults have been reported to take as much as 8 to 12 grams of L-tryptophan per day, in three to four equally divided doses. Most nutritionists and healthcare practitioners caution against taking tryptophan with protein, since protein competes with tryptophan for absorption. In contrast, vitamin B6 and magnesium support the effects of tryptophan.

Dietary Sources of Tryptophan

Foods rich in tryptophan include bananas, chocolate, cottage cheese, dried dates, fish, meat, milk, oats, peanuts and turkey. Other dietary sources with lower concentrations of tryptophan include corn, cereal grains, legumes (peas and beans), flesh foods, eggs, dairy products, some nuts and seeds and the casein component of milk. A 3-ounce serving of meat, tuna or peanuts contains about 300 milligrams of tryptophan.

Human breast milk is also a potent source of tryptophan. In fact, some studies have found that breast-fed infants have higher plasma tryptophan concentrations than formula-fed infants, despite the higher protein concentration of formulas.

Higher plasma concentrations of tryptophan may help explain the lower food intake of breast-fed infants. Because breast-fed infants and those fed formula rich in lactalbumin have higher plasma concentrations of tryptophan than of other amino acids, it is predictable that their brain tryptophan concentrations would be higher. Tryptophan is a precursor for the synthesis of serotonin, and increased synthesis and release of serotonin likely improve infants' satiety.

The Truth about the Tryptophan Scare

In 1989, a stealth outbreak of a rare, then-unrecognized blood disorder began. The disorder would eventually be called eosinophilia-myalgia syndrome, or EMS. (Eosinophils are a type of white blood cell; in cases of EMS, the eosinophil count in the blood rises.)

Initially, the illness seemed comparable to flu, but some victims suffered serious and disabling symptoms—paralysis, neurological and heart problems, myalgias (painful muscle aches), swelling of the arms and legs, cough, headaches, fatigue, memory loss, cognitive defects, cracked and swollen skin, rashes and other debilitating symptoms. Ultimately, there were 1,411 cases of EMS resulting in 19 deaths.

An intensive investigation conducted by the Centers for Disease Control and Prevention (CDC) traced the disease to a contaminated batch of L-tryptophan from a single Japanese manufacturer, Showa Denko K.K. Epidemiologic (population) studies confirmed the investigation's findings. The problem was determined to have resulted from the company's manufacturing process, which resulted in bacterial contamination, as opposed to any intrinsic problem with L-tryptophan.

Nonetheless, in view of the illness and death attributed to the tainted supplement, the U.S. Food and Drug Administration (FDA) requested an urgent recall of L-tryptophan products on November 17, 1989; 331 manufacturers, re-packers and distributors immediately complied. By February 1990, the FDA had designated tryptophan as a Class 1 substance, which indicated that the supplement posed a life-threatening health hazard, and instituted a ban on the importation of L-tryptophan. On March 22, 1990, the FDA banned the public sale of L-tryptophan completely.

For the millions of consumers who had used L-tryptophan successfully for years, and without any untoward effects, to help relieve conditions such as anxiety, depression and insomnia, the entire chain of events was disturbing. Although many L-tryptophan users sympathized with the FDA action, some wondered why a substance that had for so many years been considered safe had suddenly been banned. Others wondered why, after the source of the problem was identified and isolated, the ban was not lifted.

Some industry observers speculated that the ban was related to the introduction of the antidepressant Prozac. Up until the time of the ban,

L-tryptophan was one of the most widely used antidepressants in the United States. Was it purely coincidental that a few weeks after the ban was in place, the pharmaceutical company Eli Lilly released Prozac?

Observers also pointed out that the FDA decision to ban L-tryptophan was only a ban on consumer sales—hospitals, pharmacies, veterinarians and physicians could legally purchase injectable solutions of L-tryptophan. However, the price of L-tryptophan soared as the amino acid's status changed from over-the-counter nutritional supplement to prescription-only drug. Curiously, on February 9, 1993, a patent was granted for L-tryptophan as a prescription drug indicated for the treatment of EMS, the very illness that elicited the 1990 ban.

For Some, 5-HTP Is a Substitute

The twisted, tormented tale of tryptophan didn't just end with the new and exalted status of L-tryptophan as a prescription drug. In 1995, 5-HTP, which is just downstream from tryptophan in the manufacture of serotonin in the body, became available as a dietary supplement. 5-HTP occurs naturally in the African bean *Griffonia simplicifolia*. Because it is naturally occurring, there is no way for any company to secure a patent for 5-HTP.

5-HTP has physiological effects comparable to those of L-tryptophan. Unlike supplemental serotonin, which cannot cross the blood-brain barrier, supplemental 5-HTP enters the bloodstream and readily crosses the blood-brain barrier to the central nervous system, where it is converted into serotonin. The resulting higher levels of serotonin increase the amount of active serotonin in nerve synapses—the gaps between neurons that impulses traverse—in particular brain regions. Since a variety of conditions that decrease serotonin levels are associated with depression, it is generally accepted that normal to slightly elevated serotonin levels tend to elevate mood and prevent depression.

Still, the Return of Pure, Safe L-Tryptophan Is Heralded

In the spring of 2005, like a celebrity returning from an extended stint in rehab, L-tryptophan returned to the consumer marketplace following the lifting of import alert #54–04. Once again, L-tryptophan could be imported into the United States for use in dietary supplements. (Some contend that it never really left the consumer market since it was still imported for use in baby formulas manufactured in the United States.)

Today, safe supplies of L-tryptophan are readily available, and the nutritional supplement continues to attract healthcare professionals and consumers. Satisfied consumers report that when taken daily, L-tryptophan improves mood, promotes relaxation and sleep, relieves feelings of anxiety and nervousness, soothes symptoms of premenstrual syndrome (PMS), eases jet lag, combats depression, assists in chemical withdrawal and even helps with dieting (because it helps to reduce carbohydrate cravings).

Stress Compromises Heart Health, Immunity, and Healing

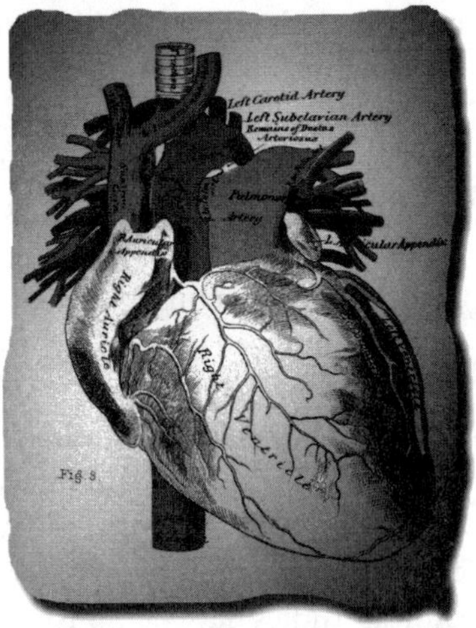

Worry and stress affects the circulation, the heart, the glands, the whole nervous system, and profoundly affects heart action.

—Charles Mayo

One of the most popular uses of L-tryptophan is as an aid to relieve stress and promote relaxation. There is a wealth of medical and scientific literature that provides evidence linking serotonin, tryptophan and 5-HTP with reduced stress and depression. We've all experienced stress at one time or another. Although it may be hard to precisely define, stress has something to do with being pushed beyond our limits. Stress is any influence that engages and challenges our physical, mental, emotional or spiritual capacity for adaptation. It is what we feel when circumstances force us beyond our natural ability to respond and adapt, challenging us to regain a sense of balance and normalcy in our lives.

The relationship between stress and cardiovascular disease is complicated. Does being a workaholic or having a Type A personality cause heart disease? Does stress raise blood pressure? Although there are many opinions about the exact nature of the relationship between stress and heart disease, research has shown that stress does contribute to heart disease in some people.

Stress may not directly cause high blood pressure, high cholesterol or other cardiac risk factors, but it can exacerbate them. What's even worse is that many of the lifestyle choices people make when they are under stress—smoking, overeating and consuming alcohol—independently increase risk for cardiovascular disease.

Scientists have known for years that major, acute stressors and chronic life stress can compromise immune function and increase susceptibility to disease. Stress increases your susceptibility to several types of infections, from the common cold to tuberculosis, as well as to autoimmune disorders, in which the body produces immune responses against its own cells.

The immune system relies on two types of white blood cells—neutrophils, the larger and more numerous white blood cells involved in fighting infection, and leukocytes, the main fighter cells. There are three kinds of leukocytes—lymphocytes, monocytes and granulocytes—each with its own unique function.

Stress-induced increases in hormone levels affect the immune system in two ways: first, they increase the quantity of neutrophils; second, they decrease the quantity of lymphocytes, which produce antibodies and other substances that fight infection and disease.

As a result of these altered levels of white blood cells, chronic stress can prolong healing and recovery from illness or injury. Laboratory and clinical studies have demonstrated that psychological stress is associated with slower healing of injuries and surgical wounds and with more complicated postoperative recovery.

The Hormonal Response to Stress

When we experience a stressful event, whether physical or mental, the endocrine system receives a message to increase its production of epinephrine (adrenaline). In response to higher levels of epinephrine, the body shifts into a tense, alert state; our breathing becomes shallow, our heart rate increases, our senses are put on high alert and we may, at least initially, become quite still, like an animal in the wild sniffing the air for danger.

The epinephrine response is very fast. Following behind this quick shift is a slower, longer-lasting increase in production of glucocorticoid hormones, primarily cortisol. The renowned stress expert Dr. Robert Sapolsky offers a colorful comparison of the relative effects of epinephrine and glucocorticoids: "In a fight-or-flight scenario, epinephrine is the one handing out guns; glucocorticoids are the ones drawing up blueprints for new aircraft carriers."

Epinephrine is event driven, shifting up and down as circumstances dictate. Cortisol is strongly tied to our circadian rhythms (daily patterns of waking and sleeping). Cortisol levels increase in the morning as we start the day and decline to their lowest point at night, several hours after we fall asleep.

Disrupting the regular hormonal pattern is one of the ways in which chronic stress wreaks havoc with our bodies. When our response to stress prevents us from synchronizing our sleep-wake cycles with our natural hormonal rhythms, the consequences can be devastating. Working to reestablish the restorative quality of sleep is one of the best things we can do for our bodies and is one of the best ways to reverse the devastating effects of stress.

Chronic Fatigue Syndrome

Chronic fatigue syndrome (CFS) is a debilitating and complicated disorder. Sufferers experience profound fatigue for six months or longer that is not relieved by bed rest and may be worsened by physical or mental exertion. People with CFS often function at much lower levels of activity than they did before they became ill. CFS also may involve other symptoms such as abdominal pain, alcohol intolerance, bloating, chest pain, chronic cough, diarrhea, dizziness, dry eyes, dry mouth, earaches, irregular heartbeat, jaw pain, morning stiffness, nausea, night sweats, psychological problems (depression, irritability, anxiety and panic attacks), shortness of breath, skin sensations, tingling sensations and weight loss.

The CDC has established two specific criteria for the diagnosis of CFS. First, patients must experience severe chronic fatigue lasting six months or longer with other known medical conditions excluded by clinical diagnosis. Second, patients must concurrently suffer from four or more of the following symptoms: substantial impairment in short-term memory or concentration; sore throat; tender lymph nodes; muscle pain; multi-joint pain without swelling or redness; headaches of a new type, pattern or severity; unrefreshing sleep; and post-exertional malaise lasting more than 24 hours.

The precise causes of CFS have not yet been identified and there are no specific tests to confirm diagnosis. In some cases, CFS can persist for years. People with CFS are sometimes treated with immune therapy, corticosteroids, antidepressants and other drugs, as well as with supplements and complementary therapies.

Research suggests that CFS causes a variety of nutritional deficiencies. For example, it is thought that tryptophan deficiency not only contributes to the symptoms of CFS but also is detrimental to the healing process. Conversely, L-tryptophan supplementation may ease the symptoms of CFS. One study found that large does of L-tryptophan (30 mg/kg) improved the ability of persons with CFS to perform cognitively demanding tasks.

Easing Depression and Enhancing Emotional Well-Being

There is a strong association between stress, anxiety, sleep loss, depression and other mental health problems. It's not surprising that all of these problems affect mood and mental health when you consider that they involve shifts in the balance of serotonin, norepinephrine, dopamine, melatonin and other neurotransmitters that govern cognition and emotional well-being.

There is also mounting evidence of a complex, multi-directional relationship between stress, sleep and health—especially mental health. Many studies have found that stress and sleep disturbances contribute to the development of various medical and psychiatric disorders or increase their severity. Research has also found that these same disorders result in increased stress and poor sleep quality. This lends credence to the hypothesis that not only do stress and sleep affect health, but also that health problems can exacerbate stress and affect sleep.

Although the relationship between stress, sleep and mental health has not yet been completely described, one possible explanation is that sleep helps us to maintain optimal emotional and social functioning by allowing the parts of the brain that control these functions to rest.

A recent review of research on the efficacy of 5-HTP, a metabolite of tryptophan, identified two well-designed published studies that found tryptophan and 5-HTP to be more effective than a placebo for mild to moderate depression.

Migraine Headaches

Practically everyone has suffered from a tension headache at one time or another. Tension headaches—characterized by aching and tightness around the forehead and temples or at the back of the head and neck—can last from a half hour to several days. Many sufferers describe the pain as feeling like a vise grip squeezing at the temples or the back of the head. Although tension headaches will subside without treatment, and most people find relief using non-prescription pain relievers (such as acetaminophen, aspirin or ibuprofen), many health-care professionals feel that stress management is the key to preventing or reducing the frequency of tension headaches.

Migraine headaches are much more serious than simple tension headaches. Migraine is a chronic neurological disease characterized by episodic attacks of excruciating, often disabling headaches with associated symptoms such as nausea, vomiting and extreme sensitivity to light and sound. Migraine headaches—often described as pulsing or throbbing—are exacerbated by physical activity and may occur on just one side of the head. An estimated 11 percent of American adults—more than 28 million Americans—suffer from migraines.

Researchers have found that tryptophan depletion increases nausea, headache and sensitivity to light in migraine sufferers. These findings suggest that a reduction in brain synthesis of serotonin intensifies light sensitivity and other migraine symptoms and may contribute to the development of migraine. Other research suggests that low levels of serotonin not only may make people susceptible to developing migraine headaches but also may be responsible for motion sickness.

While prescription drugs can help to reduce the pain and duration of migraines, many health professionals and migraine sufferers advocate prevention. Healthy living, including a varied, balanced diet, regular aerobic exercise, meditation, deep breathing and yoga, can help to reduce the frequency of migraines. In addition to biofeedback, some migraine sufferers also respond well to treatment with acupuncture or physical therapy; others benefit from supplementation with L-tryptohan, 5-HTP, feverfew, magnesium and vitamin B2.

What Causes Headaches?

Although it is often easy for healthcare practitioners to pinpoint the underlying causes of secondary headaches (those that occur in response to trauma, infection or injury or as a symptom of another disease or disorder), medical understanding of the causes of primary headaches remains incomplete. Generally, no specific cause can be named as the culprit responsible for a specific headache; sometimes, however, the anatomy of the headache—how the headache arises, how long it lasts and where the pain is felt—provides valuable clues about the origin of the headache.

Historically, scientists believed that there were two simple mechanisms underlying the development of headaches. Vascular headaches, such as migraine headaches, were thought to occur in response to dilation and constriction of the blood vessels that supply the brain and the muscles in the head and neck. Tension-type headaches were thought to result from the tensing and contracting of muscles in the head, face and neck in response to physical or emotional stress.

Recent research reveals that some headaches are associated with complex chemical changes in the brain. Tension-type headaches are associated with varying levels of serotonin, endorphins and other chemicals that serve as neurotransmitters. As previously described, serotonin is a hormone produced from tryptophan and found in the brain, blood serum and gastric mucous membranes. Serotonin acts both as a neurotransmitter and as a vasoconstrictor (a substance that causes blood vessels to narrow).

The chemical changes observed in headache sufferers may cause or be caused by muscle tension. It is possible that tense muscles in the neck and scalp trigger headaches in people with altered chemical levels; conversely, the neck and scalp muscles may tense up in response to neurochemical changes. While it's not yet known why neurotransmitter levels vary or whether their variability causes or results from muscle tension, their fluctuations are thought to not only activate pain pathways to the brain but also to interfere with the brain's ability to suppress pain.

The brain's ability to suppress pain begins in the brain stem. When the brain stem receives a pain signal, the signal is usually conveyed by the trigeminal nerve, the largest nerve in the head. When a headache trigger stimulates the trigeminal nerve, the nerve releases neurotransmitters, which in turn prompt the release of serotonin to counter the pain. Sufficient serotonin levels help to dampen the pain signals from the trigeminal nerve, but research has found that many headache sufferers do not produce sufficient serotonin.

Another neurotransmitter that may be involved in tension headaches is nitric oxide. Overproduction of nitric oxide has been linked to migraine and chronic tension headaches. Substances that block the production of nitric oxide have been shown to reduce the muscle tightness associated with tension headaches. Along with too much nitric oxide, some researchers believe that headache sufferers also have low levels of endorphins—the naturally occurring pain-relieving compounds found in the brain.

Because tension headaches and migraine headaches involve similar alterations in neurochemicals, some researchers believe that the two types of headache are related. The majority of people who suffer from migraine headaches also experience an occasional tension headache, and a quarter of people diagnosed with tension headaches experience an occasional migraine headache. Some investigators hypothesize that chronic tension headaches may predispose people to the development of migraine, with migraine symptoms emerging as the pain from tension headaches worsens. Other research posits that a mild migraine headache is a type of tension headache and that the two headaches exist along a continuum ranging from mild tension headaches to severe migraine headaches.

Premenstrual and Menstrual Problems

Premenstrual syndrome (PMS) is a condition that occurs in the last half of a woman's menstrual cycle and involves physical symptoms and mood disturbances. Physical symptoms of PMS include bloating and breast tenderness; emotional changes include feelings of anger and depression. The exact causes of PMS have not been determined; however, monthly chemical fluctuations involving sex hormones, neurotransmitters and opioid peptides may be responsible.

Many of the mood disorders associated with premenstrual syndrome and other menstrual disorders have been effectively treated using serotonin-augmenting drugs, and preliminary clinical research suggests that there may be a role for serotonin precursors such as tryptophan and 5-HTP in relieving some of these symptoms.

Interestingly, one study found that reducing serotonin by depleting tryptophan increased aggression in women during the second half of their menstrual cycles. Earlier studies had confirmed a comparable response in men—decreased serotonergic neurotransmission resulted in increased aggression in male subjects.

Smoking Cessation

There are at least two reasons that antidepressants might help people quit smoking. Antidepressants may relieve symptoms of depression associated with nicotine withdrawal. This is because nicotine may have an antidepressant effect that keeps people smoking, and antidepressants may substitute for this effect. Alternatively, some antidepressants may potentially block nicotine receptors, affecting the neural pathways underlying nicotine addiction independent of the drug's antidepressant effects.

Serotonin enhancers, such as tryptophan and high-carbohydrate diets, have been used to help smokers reduce withdrawal-related anxiety. At least one study on smokers trying to quit found that anxiety and other withdrawal symptoms were lower in subjects given tryptophan than in control subjects.

Stress, Hunger and Weight Gain

You don't get ulcers from what you eat.
You get them from what's eating you.

—Vicki Baum (1888–1960)
Austrian-American novelist

Some people may initially respond to chronic stress by losing their appetites; in the long run, however, stress leaves most people hungry and with a tendency to put on weight. This may occur because stress-disturbed metabolisms become less efficient, requiring more nutrients to produce an adequate level of cellular energy.

Weight gain also may occur as disturbed endocrine patterns block the chemical messages that communicate a sense of satisfaction and fullness. Interestingly, recent research has shown that reducing stress and improving sleep increase the action of the hormone leptin, the primary chemical messenger that tells us when we're full and directs us to stop eating.

When stress disrupts sleep patterns and depletes energy levels, people naturally seek stimulants such as the caffeine in coffee and soft drinks. People may also seek a quick energy boost from sugars and other simple carbohydrates found in bread, cookies and candy. These simple carbohydrates rapidly enter the bloodstream, temporarily boosting energy levels.

But this boost comes at a price. Stimulants chemically quicken the sympathetic autonomic responses. This creates more stress and causes the cells to run "hotter," demanding more nutrients and producing more toxic waste products. By forcing our bodies to churn out more epinephrine, we tend to desensitize our epinephrine receptors, setting up a vicious cycle in which we need more stimulation to achieve the same effects.

There are other problems with this pattern. Fat cells need to receive signals from the nervous system to be broken down to produce energy. But the more we desensitize our epinephrine receptors, the harder it becomes for those signals to get through. Fat then accumulates and it becomes extremely difficult to lose weight, even when we diet and engage in vigorous exercise—the fat cells simply can't hear the call to break down.

When we eat simple carbohydrates, insulin levels spike. Chronically high levels of insulin make cells shut down their insulin response, leading to impaired glucose tolerance, which may progress to type 2 diabetes. As is the case with stimulants, we find ourselves needing more and more sugar and high-carbohydrate foods to get a wake-up call.

Obviously, extra carbohydrates are fattening. And excess insulin doesn't just hang around. Although its primary function is to chaperone glucose molecules into our cells to be burned as fuel, a secondary function of insulin is to direct the storage of extra calories as fat—the very same fat that won't break down because the fat cells have become epinephrine resistant.

Stress—even mental stress—quickly affects core physiological systems in a complex and aggravating fashion. Recovering from the effects of stress, especially chronic stress, is difficult, but not impossible. Because our body's responses to stress span so many systems, the path back to health must effectively address a range of physical, emotional and nutritional needs.

In obese persons, plasma tryptophan concentrations have been shown to be decreased, independent of weight and diet. Decreased tryptophan levels may reduce serotonin production and cause mood disturbances, depression and impaired satiety, all of which may ultimately lead to increased caloric uptake and obesity.

Research also reveals that the ratios of tryptophan to other amino acids (plasma tryptophan ratios) are low in obese people. Plasma tryptophan ratios predict brain tryptophan uptake and serotonin production. Considerable research suggests that L-tryptophan and 5-HTP promote weight loss by enhancing feelings of satiety and helping to control carbohydrate cravings.

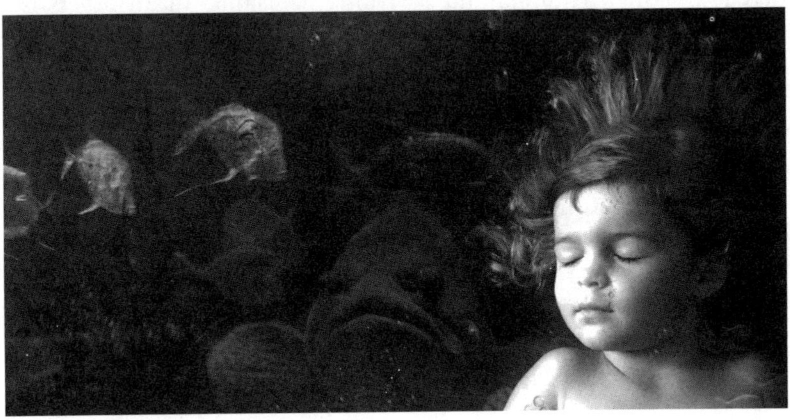

The Miracle of Sleep

For fast-acting relief, try slowing down.

—Lily Tomlin

If it's not a problem in our lives, we tend to take sleep for granted. But poor sleep patterns can create a host of problems. People who don't get enough sleep, or whose sleep is not refreshing, are frequently at risk of experiencing episodes of "micro-sleep"—blackouts that they may not remember. Obviously, these episodes can be extremely dangerous, especially when they occur while a person is driving or performing other activities that require constant attention.

Sleep deprivation can also cause more subtle problems. Sleep is a time when the body regenerates and detoxifies. When sleep is disturbed, the stage is set for problems ranging from accelerated aging, disruptive variations in mood and attention, increased susceptibility to infections and other immune system problems.

Every night, we enter a special stage of sleep called REM (rapid eye movement) sleep. REM sleep is the type of sleep during which we dream (whether or not we consciously remember our dreams). Some neurologists and sleep researchers believe that REM sleep is an essential period in which the brain reorganizes memory and experience.

Sleep deprivation experiments show that individuals who are continually blocked from entering REM sleep undergo rapid mental deterioration. Within a few days, they may even start to exhibit signs of severe mental disorders including auditory and visual hallucinations, disordered thinking, paranoia and psychotic delusions. In an otherwise healthy person, these symptoms are temporary and quickly recede when REM sleep is once again permitted.

When stress derails and degrades neurological and hormonal relationships with sleep, people may begin to exhibit the symptoms of sleep deprivation in more subtle ways, putting a tremendous strain on work, personal relationships and self-esteem. As with other physiological effects of stress, the degradation of sleep is a vicious cycle. Stress interferes with sleep, and lack of sleep increases stress. And the impact of this spiraling problem, through its effects on mood, attention and cognitive function, can cause problems that may become new stress factors. The importance of restful sleep cannot be overstated.

The Chemistry of Sleep and Wakefulness

Neurotransmitters, the chemicals in the brain that transmit messages from one nerve cell to another, are involved in regulating sleep and wakefulness. Neurons in the brain stem produce neurotransmitters such as serotonin and norepinephrine that keep parts of the brain active while we are awake. Other neurons at the base of the brain begin signaling when we fall asleep and dampen or turn off the signals that keep us awake.

Research suggests that while we are awake a sleep-producing substance called adenosine accumulates in the brain and in the blood, causing drowsiness. Adenosine is a breakdown product of adenosine triphosphate, or ATP, an important source of cellular energy. While we sleep, adenosine gradually breaks down. Research on animals has found that adenosine levels in the brain rise during sleep deprivation and return to baseline during sleep. The caffeine in your eye-opening morning cup of coffee is thought to promote wakefulness by blocking adenosine receptors.

The part of the brain that is most important in regulating sleep is the hypothalamus. The role of the hypothalamus in regulating sleep was recognized after the worldwide spread of encephalitis lethargica in the early 1900s. Thought to be a viral infection, encephalitis lethargica causes severe sleep disturbances in infected persons. During the previously mentioned outbreak, most patients became profoundly sleepy but others suffered from insomnia. The patients afflicted with sleepiness were found to have lesions in the posterior hypothalamus, while those with insomnia had lesions in the anterior hypothalamus.

The anterior hypothalamus contains groups of hypothalamic neurons that produce gamma-aminobutyric acid (GABA). GABA is a neurotransmitter that promotes sleep by inhibiting neurons involved in wakefulness, including neurons containing histamine, norepinephrine, serotonin, hypocretin and glutamate.

Hypocretin (also called orexin) was discovered in 1998, and its role in sleep was identified in 2001. The hypocretin-producing neurons in the hypothalamus receive direct input from the suprachiasmatic nucleus (SCN) and, in accord with circadian rhythm, their concentration is highest during wakefulness. Hypocretin levels also increase during periods of forced sleep deprivation. Other as-yet-undiscovered transmitters are undoubtedly involved in maintaining wakefulness and promoting sleep .

The Serenity of Serotonin

When nerve cells talk to one another using electrical impulses, neurotransmitters shape and direct their communication by amplifying, relaying, blocking or modulating the messages. Serotonin is one of the most important neurotransmitters and has been identified as a key regulator of mood, sleep, sexuality and appetite. Many antidepressant drugs—selective serotonin reuptake inhibitors, or SSRIs—work by blocking the body's resorption of serotonin, effectively keeping each molecule in play for a longer period of time.

It's tempting to think that we can relieve our stress, anxiety or depression by taking serotonin in the form of a pill. Unfortunately, serotonin does not pass through the blood-brain barrier, and any serotonin we consume will be broken down without doing any us good. In fact, when injected under the skin, serotonin acts as a painful irritant—the venom of bees and wasps actually contains serotonin!

Serotonin levels can, however, be influenced by consuming the precursors from which the body makes serotonin. Serotonin is synthesized from the amino acid L-tryptophan, which is then converted into 5-HTP, an intermediate chemical form. Many people report that taking L-tryptophan or 5-HTP brightens their mood, improves the quality of their sleep, helps them to resist depression and anxiety and relieves many symptoms of stress.

In the body, 5-HTP passes through the blood-brain barrier into the central nervous system, where it is converted into serotonin. While this nutritional approach works well for many people, the relationships between neurotransmitter levels, mood and behavior are very complex and highly individual. Experiment with supplemental 5-HTP to find out if it works for you.

. . . and Melatonin

Melatonin is a neurohormone produced from tryptophan by the pineal gland (located in the brain). Its synthesis and release are stimulated by darkness and suppressed by light, indicating that melatonin plays an important role in circadian rhythms. Blood levels of melatonin are elevated just before bedtime, and peak levels occur from 2–4 A.M. Melatonin supplements have been used to help travelers recover from jet lag and to treat insomnia and other sleep disorders.

Because melatonin is so popular among people suffering from sleep disorders, the National Center for Complementary and Alternative Medicine funded research that analyzed available scientific evidence supporting its safety, efficacy and risks. The researchers found that melatonin seems safe for short-term use but may not be effective for treating sleep disorders such as jet lag. On the other hand, the researchers noted that melatonin shows promise for delayed sleep phase syndrome. Melatonin's precise mechanism of action has not yet been uncovered and further research is needed to determine its best use in treating sleep disorders.

References

Anderson GH. "Regulation of food intake." In *Modern nutrition in health and disease,* Shils M, Olson J, Shike M, eds. Malvern, Lea and Febiger. 1994: 524–36.

Bell SJ, Goodrick GK. "A functional food product for the management of weight." *Crit Rev Food Sci Nutr.* 2002 Mar; 42(2): 163–78.

Bond AJ, Wingrove J, Critchlow DG. "Tryptophan depletion increases aggression in women during the premenstrual phase." *Psychopharmacology* (Berl). 2001 Aug; 156(4): 477–80.

Bowen DJ, Spring B, Fox E. "Tryptophan and high-carbohydrate diets as adjuncts to smoking cessation therapy." *J Behav Med.* 1991 Apr; 14(2): 97–110.

Brandacher G, Hoeller E, Fuchs D, Weiss HG. "Chronic immune activation underlies morbid obesity: is IDO a key player?" *Curr Drug Metab.* 2007 Apr; 8(3): 289–95.

Breum L, Rasmussen MH, Hilsted J, Fernstrom JD. "Twenty-four-hour plasma tryptophan concentrations and ratios are below normal in obese subjects and are not normalized by substantial weight reduction." *Am J Clin Nutr.* 2003 May; 77(5): 1112–18.

Brey RL. "Both migraine and motion sickness may be due to low brain levels of serotonin." *Neurology.* 2005 Aug 23; 65(4): E9–10.

Carter HE, Con MJ. "William Cumming Rose. Biographical Memoirs." National Academy of Sciences. http://books.nap.edu/html/biomems/wrose.html.

"Chronic Fatigue Syndrome." Centers for Disease Control and Prevention. http://www.cdc.gov/cfs/cfsdefinition.htm. May 3, 2006.

Drummond PD. "Tryptophan depletion increases nausea, headache and photophobia in migraine sufferers." *Cephalalgia.* 2006 Oct; 26(10): 1225–33.

Eriksson O, Wall A, Marteinsdottir I, Agren H, Hartvig P, Blomqvist G, Langstrom B, Naessen T. "Mood changes correlate to changes in brain serotonin precursor trapping in women with premenstrual dysphoria." *Psychiatry Res.* 2006 Mar 31; 146(2): 107–16.

Lien EL. "Infant formulas with increased concentrations of alpha-lactalbumin." *Am J Clin Nutr.* 2003; 77: S1555–58.

Manders DW. "The FDA Ban of L-Tryptophan: Politics, Profits and Prozac." Life Extension. http://www.lef.org/fda/fdaban95.html. April 6, 1998.

Markov D, Goldman M. "Normal Sleep and circadian Rhythms: Neurobiological Mechanisms Underlying Sleep and Wakefulness." *Psychiatr Clin N Am.* 2006; 29: 841–53.

Morgan RM, Parry AM, Arida RM, Matthews PM, Davies B, Castell LM. "Effects of elevated plasma tryptophan on brain activation associated with the Stroop task." *Psychopharmacology* (Berl). 2007 Feb; 190(3): 383–89.

Richell RA, Deakin JF, Anderson IM. "Effect of acute tryptophan depletion on the response to controllable and uncontrollable noise stress." *Biol Psychiatry.* 2005 Feb 1; 57(3): 295–300.

Sapolsky R, "Taming Stress." *Scientific American.* 2003 Sept.

Shaw K, Turner J, Del Mar C. "Tryptophan and 5-hydroxytrytophan for depression." *Cochrane Database Syst Rev.* 2002; (1): CD003198.

Werbach MR. "Nutritional strategies for treating chronic fatigue syndrome." *Altern Med Rev.* 2000 Apr; 5(2): 93–108.

Wilcox LS. "Old black water." *Prev Chronic Dis* [serial online] Nov 2005. http://www.cdc.gov/pcd/issues/2005/?nov/05_0191.htm.

Zee PC, Turek FW. "Sleep and health: Everywhere and in both directions." *Arch Intern Med.* 2006; 166(16): 1686–88.